Early Childhood Technology Planning

Susan Louise Peterson

Early Childhood Technology Planning
Susan Louise Peterson

COPYRIGHT

Copyright 2015 by Susan Louise Peterson

CONTENTS

Preface vii

Prologue ix

Acknowledgements xi

Introduction xiii

Chapter 1: What is Early Childhood Technology? 1

Chapter 2: Early Childhood Technology Teaching Strategies 11

Chapter 3: Early Childhood Technology Rewards 53

Chapter 4: Safety And Early Childhood Technology 63

Chapter 5: Questions About Early Childhood Technology 73

Recommended Reading 93

Index 95

Afterword 101

PREFACE

Early Childhood Technology Planning is a book written to help teachers and librarians plan educational programs in early childhood technology. My motivation to write this planning guide was from seeing the lack of educational materials on the topic of technology for young learners. My experience as an early childhood educator affords me to know that many teachers of young children have very little time to write about technology because they are so busy in their classrooms, preschools and day care centers. Children's librarians face the same problems as they are busy with story time and developing library programs for young children. There is a great need to make connections between early childhood programs that are so successful and the technology that is important for the future of young children. The purpose of *Early Childhood Technology Planning* is to give assistance to early childhood professionals as they plan and organize their early childhood education programs with a technology component.

PROLOGUE

This search for using technology in the early childhood setting is certainly one that requires planning and forethought. The early childhood educator must think through everything from the rationale for using early childhood technology to the resources available in the school, classroom or library. This planning may involve using various technology teaching strategies from many angles and perspectives. Rewards and praise can certainly go a long way for encouraging the confidence of young children in using technological information. Of course, every early childhood professional must recognize the importance of providing a safe technology environment in the early childhood classroom or library.

Early Childhood Technology Planning is written to help in the beginning phase of planning for the early childhood educator and librarian.

ACKNOWLEDGEMENTS

I want to thank the many early childhood professionals and early childhood librarians who have given me so many ideas about using technology in early childhood settings. The examples these professionals provided for me were engaging and creative in working with my early childhood students.

I also want to thank my husband (my best tech support ever) and my beautiful daughters for their continued support in my writing hobby, passion and career. I hope they each find their passion in life (as I have found mine).

INTRODUCTION

Early Childhood Technology Planning is geared as a planning book to help early childhood educators and librarians' booster the technology learning environment for young children. Not only does the book help early childhood professionals utilize resources, but understand the challenges faced when using early childhood technology. Technology in the early childhood setting is totally revolved around the discovery approach where young children are using their curiosity and senses to explore materials and organize information with a technological approach. Praise and motivation are equally important to encourage positive technological experiences for young children. In any technology environment there will always be a safety supervisory component where caution and rules are promoted to prevent accidents and encourage a safety in the early childhood setting. The book ends with questions early childhood professionals have about technology to encourage future technology planning topics.

CHAPTER 1

WHAT IS EARLY CHILDHOOD TECHNOLOGY?

This book is written to explore the topic of early childhood technology. The rapidly changing world of technology must address the needs of young children to prepare them for a future filled with technological experiences. Teachers, librarians, parents and professionals working with young children must plan effective programs and experiences to enhance the learning of young children. The goal of early childhood technology is to prepare young children to face a broad range of future technological situations and experiences. For the purpose of this book early childhood technology is defined below:

EARLY CHILDHOOD TECHNOLOGY

Early childhood technology refers to planned educational experiences to introduce young children to technology situations and to encourage positive technological experiences.

Professionals working with early learners have much to do in planning technology activities and developing the goals of an early childhood technology program. The work will be demanding, but it can be rewarding to see young children grow and obtain positive self-assurance about the world of technology.

RATIONALE FOR EARLY CHILDHOOD TECHNOLOGY

1. Some young children lack technological experiences at home, so schools, day care facilities and libraries may be the only place they receive the training.
2. Young children need to be introduced to technology early, so they can build on their technology skills and continue through the upper grades and college levels.
3. Young children will need technology skills for future employment opportunities.
4. Rapidly changing technology requires that children be flexible and adaptable to technological advancements.
5. Many future learning opportunities will be technology based, so technology experiences must be introduced early to children.
6. Our society is becoming information connected using technology. Children will need technology skills to communicate effectively in this information-oriented society.
7. A young child needs technology opportunities to develop a positive self-image and have confidence in using computers and technological equipment.
8. New early childhood technology experiences can enhance and enrich an established curriculum.

OBJECTIVES OF EARLY CHILDHOOD TECHNOLOGY

The objectives for early childhood technology include the following:

- To introduce early learners to beginning technology vocabulary.
- To provide young children opportunities to get hands on experience with computers and technological equipment.
- To show early learners how they can work cooperatively with others on technology projects.
- To demonstrate to young children how technology is used in our world.
- To help children complete simple tasks on computers and technological equipment.
- To assist young children in gaining positive technology learning experiences.
- To provide information to young children on establishing good safety habits when using technological devices.
- To give young children an appreciation of the role technology plays in society.

CHALLENGES OF EARLY CHILDHOOD TECHNOLOGY

The challenges of early childhood technology include:

- Limited technology training for early childhood professionals
- Lack of technological equipment in early childhood facilities and libraries
- Uncooperative administrators and managers may not support the idea of early childhood technology
- Limitations of technology opportunities may exist from class size or behavioral difficulties
- Lack of consistency in early childhood technology programs in a large district or region
- Limited technological support for the child from the home environment
- Lack of funding for technological equipment and supplies
- Lack of awareness of available technology resources

EARLY CHILDHOOD TECHNOLOGY RESOURCES

New early childhood technology resources are popping up everyday. Here is a list of some of these resources:

INTERNET EDUCATIONAL INFORMATION
LEARNING WEBSITES
INSTRUCTIONAL SOFTWARE PROGRAMS
KIDS WEBCAMPS
INTERNET /ONLINE SCHOOLS
ONLINE EDUCATIONAL CATALOGS
COMPUTER CAMPS
TECHNOLOGY EDUCATIONAL & APPLICATION PROGRAMS
TECHNOLOGY TRAINING WORKSHOPS
INDIVIDUALIZED COMPUTER TRAINING

EARLY CHILDHOOD TECHNOLOGY INFORMATION

There is a tremendous amount of early childhood education technology information available to parents, librarians and teachers. The following is a list of some of this technology information:

CHILDREN'S MUSEUM INFORMATION
CHILDREN'S SAFETY INFORMATION
EDUCATIONAL TOY INFORMATION
LEARNING MATERIALS INFORMATION
PARK AND RECREATION INFORMATION
AMUSEMENT ATTRACTION INFORMATION
SIGHTSEEING INFORMATION
MUSIC AND GAME INFORMATION
ANIMAL DESCRIPTION INFORMATION

TEN WAYS TEACHERS CAN USE EARLY CHILDHOOD TECHNOLOGY INFORMATION

PLANNING FIELD TRIPS

DEVELOPING LESSON PLANS

IMPLEMENTING CURRICULUM

CONDUCTING PARENT MEETINGS

MAKING HOME VISITS

DEVELOPING TEACHER INSERVICE TRAINING

PLANNING TECHNOLOGY WORKSHOPS

CREATING NEW CLASSROOM ACTIVITIES

INTRODUCING STUDENTS TO TECHNOLOGY

PROVIDING PLAY ACTIVITIES WITH TECHNOLOGY

TEN WAYS LIBRARIANS CAN USE EARLY CHILDHOOD TECHNOLOGY INFORMATION

PREPARING WEEKLY STORY TIME

DEVELOPING CHILDREN'S READING ACTIVITIES

PLANNING READING EVENTS

CREATING COMMUNITY OUTREACH PROGRAMS

ARRANGING LIBRARY AND SCHOOL PARTNERSHIPS

PRESENTING LIBRARY TRAINING PROGRAMS

TEACHING READING WORKSHOPS

PROMOTING EXCITEMENT ABOUT READING

CONDUCTING SUMMER READING PROGRAMS

PROVIDING READING AND TECHNOLOGY ENRICHMENT

TEN WAYS PARENTS CAN USE EARLY CHILDHOOD TECHNOLOGY INFORMATION

PLANNING VACATIONS

PURCHASING EDUCATIONAL SOFTWARE

ANSWERING CHILDREN'S HEALTH QUESTIONS

DEVELOPING A SPECIALIZED LEARNING PROGRAM

ARRANGING A TUTORIAL PROGRAM

STRUCTURING A FAMILY EVENT

CONDUCTING A FAMILY PROJECT

SCHEDULING FAMILY ACTIVITIES

REVIEWING CHILDREN'S CAMPS

INTRODUCING NEW SKILLS TO CHILD

CHAPTER 2

EARLY CHILDHOOD TECHNOLOGY TEACHING STRATEGIES

This chapter focuses on some teaching strategies early childhood teachers and librarians can use in helping children with technology. A number of teaching strategies can be used to help young children have a variety of experiences with technology. Professionals working with young children must use a number of approaches because these children have different learning styles. In addition, young children are exploring and discovering their learning skills at the same time they are being introduced to technology. Teaching strategies can easily be applied to the early childhood areas of:

PLANNING EXPERIENCES
FINE MOTOR EXPERIENCES
GROSS MOTOR EXPERIENCES
SENSORY EXPERIENCES
RESPONSIBILITY EXPERIENCES
ART EXPERIENCES
DISCOVERY EXPERIENCES
ROLE PLAYING EXPERIENCES
CLASSIFICATION EXPERIENCES

ENRICHMENT EXPERIENCES
SPATIAL RELATIONS EXPERIENCES
SOCIAL EXPERIENCES
TIME EXPERIENCES
AUDITORY EXPERIENCES
LANGUAGE EXPERIENCES
VISUAL EXPERIENCES
CREATIVE EXPERIENCES
EMOTION EXPERIENCES
MUSIC EXPERIENCES
POSITIVE LEARNING EXPERIENCES

EARLY CHILDHOOD TECHNOLOGY TEACHING STRATEGY #1

PLANNED TECHNOLOGY EXPERIENCES
The early childhood professional should provide planned technology experiences for young children.

TEACHING STRATEGIES TO ENCOURAGE PLANNED TECHNOLOGY EXPERIENCES

- PLAN A YEARLY TECHNOLOGY CALENDAR OF ACTIVITIES
- COORDINATE TECHNOLOGY EXPERIENCES WITH SCHOOL AND LIBRARY EVENTS
- MAKE A MONTHLY SCHEDULE OF TECHNOLOGY EVENTS
- PROVIDE YOUNG CHILDREN WITH CHOICES OF TECHNOLOGY ACTIVITIES
- DEVELOPING TECHNOLOGY EVENTS FOR YOUNG CHILDREN TO DISPLAY THEIR WORK
- HELP YOUNG CHILDREN HAVING DIFFICULTY BY PROVIDING REMEDIAL TECHNOLOGY ACTIVITIES
- GIVE YOUNG CHILDREN OPPORTUNITIES TO DISPLAY THEIR TECHNOLOGY SKILLS
- FURNISH YOUNG CHILDREN WITH PLENTY OF TECHNOLOGY MATERIALS AND SUPPLIES

- PREPARE YOUNG CHILDREN FOR TECHNOLOGY ACTIVITIES BY PROVIDING SPECIFIC DIRECTIONS
- DEMONSTRATE HOW TECHNOLOGY EQUIPMENT WORKS BEFORE YOUNG CHILDREN BEGIN TECHNOLOGY LESSONS

EARLY CHILDHOOD TECHNOLOGY TEACHING STRATEGY #2

TECHNOLOGY FINE MOTOR EXPERIENCES

The early childhood professional will help young children develop fine motors experiences with the use of technology.

TEACHING STRATEGIES TO ENCOURAGE FINE MOTOR EXPERIENCES

- HAVE YOUNG CHILDREN EXERCISE FINGERS BEFORE WORKING WITH TECHNOLOGY EQUIPMENT
- HAVE YOUNG CHILDREN PRACTICE HOLDING AND USING THE MOUSE ON THE COMPUTER
- HAVE YOUNG CHILDREN USE A PENCIL TO DRAW A PICTURE OF SOMETHING THEY SEE ON A COMPUTER
- HAVE YOUNG CHILDREN PRACTICE CUTTING OUT A COMPUTER GENERATED PICTURE
- HAVE YOUNG CHILDREN USE A FINGER TO TRACE A PICTURE ON THE COMPUTER SCREEN
- HAVE YOUNG CHILDREN REACH AND GRASP COMPUTER DISKS WITH THEIR FINGERS AND HANDS
- HAVE YOUNG CHILDREN FINISH AN INCOMPLETE PICTURE ON A COMPUTER

- HAVE YOUNG CHILDREN USE A CRAYON TO COPY A PICTURE FROM A COMPUTER
- HAVE YOUNG CHILDREN PRACTICE POINTING TO LETTERS ON A KEYBOARD
- HAVE YOUNG CHILDREN TRACE THE DOTTED LINE OF A COMPUTER GENERATED PICTURE

EARLY CHILDHOOD TECHNOLOGY TEACHING STRATEGY #3

TECHNOLOGY GROSS MOTOR EXPERIENCES

The early childhood professional will help young children develop gross motor experiences in a technology center.

TEACHING STRATEGIES TO ENCOURAGE TECHNOLOGY GROSS MOTOR EXPERIENCES

- LET YOUNG CHILDREN PARTICIPATE IN STRETCHING ACTIVITIES BEFORE USING TECHNOLOGY EQUIPMENT
- HAVE YOUNG CHILDREN IMITATE A BODY MOVEMENT THEY OBSERVE ON A COMPUTER
- ASK YOUNG CHILDREN TO SKIP TO THE TECHNOLOGY CENTER
- GIVE YOUNG CHILDREN THE OPPORTUNITY TO IDENTIFY PICTURES OF MOTOR ACTIVITIES (LIKE RUNNING AND JUMPING) SHOWN ON A COMPUTER
- PLACE A LONG PIECE OF TAPE ON THE FLOOR AND LET YOUNG CHILDREN BALANCE TO THE TECHNOLOGY CENTER

- HAVE ALL YOUNG CHILDREN IN THE TECHNOLOGY GROUP HOP UP AND DOWN FIVE TIMES
- HAVE YOUNG CHILDREN IN THE TECHNOLOGY GROUP PRACTICE TEN JUMPING JACKS BEFORE THEY START TO WORK
- WHEN YOUNG CHILDREN COMPLETE THE TECHNOLOGY ACTIVITY, LET THEM WALK SIDEWAYS TO THE NEXT CENTER
- PROVIDE YOUNG CHILDREN A MAT TO EXERCISE ON WHILE WAITING THEIR TURN TO USE THE COMPUTER
- GIVE YOUNG CHILDREN A CHANCE TO RUN IN A RACE WHEN THEY COMPLETE THE COMPUTER ACTIVITY

EARLY CHILDHOOD TECHNOLOGY TEACHING STRATEGY #4

TECHNOLOGY SENSORY EXPERIENCES

The early childhood professional will assist young children in discovering their senses through technology experiences.

TEACHING STRATEGIES TO ENCOURAGE TECHNOLOGY SENSORY EXPERIENCES

- ALLOW YOUNG CHILDREN TO FEEL DIFFERENT TEXTURES RELATED TO A TECHNOLOGY PROGRAM
- PROVIDE YOUNG CHILDREN WITH LETTERS OF THE ALPHABET TO TRACE WITH THEIR FINGERS AS THEY VIEW THE LETTERS ON THE SCREEN
- AFTER A SOFTWARE PROGRAM HAS BEEN VIEWED BY THE CHILDREN, LET THEM PAINT A PICTURE OF IT
- ASK YOUNG CHILDREN TO COMPARE HOW THE MOUSE PAD AND THE MONITORS FEEL (HARD AND SOFT)
- LET THE YOUNG CHILDREN HEAR SOUNDS OF ANIMALS WHILE LOOKING AT AN ANIMAL SOFTWARE PROGRAM
- COOK SOMETHING (I.E. POPCORN) FOR YOUNG CHILDREN TO SMELL WHILE WORKING AT THE COMPUTER. HAVE A TREAT WHEN THE ACTIVITY IS FINISHED
- HAVE CHILDREN MAKE A COLLAGE FROM THE ACTUAL

ITEMS THEY PREVIOUSLY VIEWED ON THE COMPUTER
- LET CHILDREN LOOK AT PICTURES THAT ARE ON THE SAME TOPIC AS THE SOFTWARE PROGRAM
- GIVE CHILDREN THE OPPORTUNITY TO HOLD A STUFFED ANIMAL WHILE WAITING THEIR TURN TO USE THE TECHNOLOGY EQUIPMENT

EARLY CHILDHOOD TECHNOLOGY TEACHING STRATEGY #5

TECHNOLOGY RESPONSIBILITY EXPERIENCES

The early childhood professional will provide technology experiences for growing children to learn about responsibility.

TEACHING STRATEGIES TO ENCOURAGE TECHNOLOGY RESPONSIBILITY EXPERIENCES

- PROVIDE OPPORTUNITIES FOR STUDENTS TO PICK THE SOFTWARE PROGRAM THEY WANT TO USE FOR A DAY
- GIVE CHILDREN THE CHANCE TO PICK THEIR SEAT AT THE COMPUTER TABLE
- ENCOURAGE YOUNG CHILDREN TO PUT THE COMPUTER PROGRAM IN A FOLDER AFTER THEY USE IT
- MAKE TECHNOLOGY ASSIGNMENTS WHERE YOUNG CHILDREN CAN WORK AT THEIR OWN PACE
- LET YOUNG CHILDREN HAVE A CHOICE IN THE TOPICS THEY WANT TO STUDY ON THE COMPUTER
- PICK ONE CHILD TO BE THE DAILY TECHNOLOGY LEADER AND COLLECT THE FILES AND ASSIGNMENTS

- ASK YOUNG CHILDREN DECISION MAKING QUESTIONS TO SEEK THEIR OPINIONS
- GIVE THE CHILDREN THE OPPORTUNITY TO PASS OUT MATERIALS FOR THE TECHNOLOGY CENTER
- LET YOUNG CHILDREN HELP MOVE THINGS AROUND AND ORGANIZE THE TECHNOLOGY CENTER
- ENCOURAGE CHILDREN TO PICK UP TRASH WHEN FINISHED IN THE TECHNOLOGY CENTER

EARLY CHILDHOOD TECHNOLOGY TEACHING STRATEGY #6

TECHNOLOGY ART EXPERIENCES

The early childhood professional will encourage children to learn about art through technology experiences.

TEACHING STRATEGIES TO ENCOURAGE TECHNOLOGY ART EXPERIENCES

- LET YOUNG CHILDREN FEEL A VARIETY OF ART SURFACES AFTER VIEWING THE OBJECT ON A SCREEN
- USE TECHNOLOGY TO SHOW YOUNG CHILDREN HOW ART IS USED IN SOCIETY
- PROVIDE ART MATERIALS FOR CHILDREN TO EXPERIMENT FOLLOWING A TECHNOLOGY PRESENTATION
- FOLLOWING A TECHNOLOGY LESSON, LET CHILDREN PAINT A PICTURE OF A SHAPE OR OBJECT
- SHOW YOUNG CHILDREN HOW ART IS DEVELOPED THROUGH DIFFERENT CULTURES WITH TECHNOLOGY
- LET YOUNG CHILDREN DESIGN AND CREATE PICTURES TO ILLUSTRATE SOMETHING THEY OBSERVE IN A COMPUTER PROGRAM OR APPLICATION
- ENCOURAGE YOUNG CHILDREN TO USE CLAY AND

MAKE THE CHARACTERS FROM AN APPLICATION OR COMPUTER PROGRAM
- PROVIDE CRAYONS AND MARKERS TO ENCOURAGE YOUNG CHILDREN TO DRAW SOMETHING THEY VIEWED ON A VIDEO
- LET THE YOUNG CHILDREN USE CHALK BOARDS TO DRAW AND ERASE PICTURES FOLLOWING A TECHNOLOGY LESSON
- MAKE COLLAGES TO LET CHILDREN FEEL DIFFERENT TEXTURES OF THE ITEMS THEY COLLECTED RELATED TO A VIDEO PRESENTATION

EARLY CHILDHOOD TECHNOLOGY TEACHING STRATEGY #7

TECHNOLOGY DISCOVERY EXPERIENCES

The early childhood professional will encourage young children to learn through discovery when using technology.

TEACHING STRATEGIES TO ENCOURAGE TECHNOLOGY DISCOVERY EXPERIENCES

- SHOW YOUNG CHILDREN A TECHNOLOGY ACTIVITY AND LET THEM GATHER THE ACTUAL ITEMS IN THE ROOM OR CENTER
- TAKE YOUNG CHILDREN ON A WALK AROUND THE SCHOOL OR LIBRARY TO DISCOVER THINGS THAT WERE INTRODUCED THROUGH SOFTWARE PROGRAMS
- PROVIDE CHILDREN BRAINSTORMING ACTIVITIES RELATED TO THE LESSON THEME AND LET THEM EXPLORE TECHNOLOGY
- ASK INVESTIGATIVE QUESTIONS TO ENCOURAGE YOUNG CHILDREN TO SEEK NEW INFORMATION WITH TECHNOLOGY
- ADD A FIELD THAT BRINGS TO LIFE THE TECHNOLOGY INTRODUCED TO CHILDREN
- ENCOURAGE YOUNG CHILDREN TO FIND BOOKS IN THE

ROOM WITH THE SAME THEME OR PICTURE OF THE TECHNOLOGY LESSON
- READ YOUNG CHILDREN STORIES AND HAVE THEM DISCOVER THE SIMILARITIES AND DIFFERENCES WITH A SOFTWARE PROGRAM
- LET YOUNG CHILDREN LISTEN TO A SONG AND IDENTIFY SIMILAR OBJECTS TO THAT FROM THE COMPUTER PROGRAM OR APPLICATION
- GIVE YOUNG CHILDREN THE OPPORTUNITY TO FIND LETTERS AND WORDS IN THE ROOM FOLLOWING A TECHNOLOGY ACTIVITY
- USE TECHNOLOGY TO TEACH YOUNG CHILDREN ABOUT FOOD AND THEN HAVE A TASTING PARTY AS A FOLLOW UP ACTIVITY

EARLY CHILDHOOD TECHNOLOGY TEACHING STRATEGY #8

TECHNOLOGY ROLE PLAYING EXPERIENCES

The early childhood professional will provide technology experiences for young children to develop role-playing skills.

TEACHING STRATEGIES TO ENCOURAGE TECHNOLOGY ROLE PLAYING EXPERIENCES

- PROVIDE YOUNG CHILDREN WITH CLOTHES TO DRESS UP AND ROLE PLAY CHARACTERS FROM COMPUTER PROGRAMS
- HAVE YOUNG CHILDREN IMITATE SOUNDS OF VEHICLES THEY HEARD IN A TECHNOLOGY ACTIVITY
- GIVE YOUNG CHILDREN THE OPPORTUNITY TO DRAMATIZE A STORY THEY PREVIOUSLY VIEWED FROM A MEDIA PRESENTATION
- USE POEMS AND STORIES TO ROLE PLAY A VARIETY OF EVENTS FOLLOWING TECHNOLOGY ACTIVITIES
- LET YOUNG CHILDREN PRETEND TO BE AN ANIMAL THEY OBSERVED ON THE COMPUTER SCREEN

- BRING TOYS AND OBJECTS SO YOUNG CHILDREN CAN PLAY AND RELATE ACTUAL OBJECTS FOR THE THINGS THEY OBSERVED ON THE SCREEN
- GIVE YOUNG CHILDREN THE CHANCE TO TELL STORIES OF WHAT THEY EXAMINED IN SOFTWARE PROGRAMS
- HAVE A BOX OF PROPS SO YOUNG CHILDREN CAN MAKE UP SKITS FOLLOWING TECHNOLOGY ACTIVITIES
- USE THE TECHNOLOGY ACTIVITY SO YOUNG CHILDREN CAN EXPRESS THEIR FEELINGS AND IDEAS
- LET YOUNG CHILDREN ROLE PLAY A FAVORITE PART OF A SHOW THEY OBSERVED DURING A TECHNOLOGY PRESENTATION

EARLY CHILDHOOD TECHNOLOGY TEACHING STRATEGIES #9

TECHNOLOGY CLASSIFICATION EXPERIENCES

The early childhood professional will encourage children to gain classification experiences with technology use.

TEACHING STRATEGIES TO ENCOURAGE TECHNOLOGY CLASSIFICATION EXPERIENCES

- PROVIDE YOUNG CHILDREN WITH SORTING EXPERIENCES FOLLOWING A TECHNOLOGY ACTIVITY
- LET YOUNG CHILDREN MATCH COLORS AND SHAPES INTO CATEGORIES FROM SOFTWARE PROGRAMS
- SHOW YOUNG CHILDREN OBJECTS WITH TWO CHARACTERISTICS (I.E. RED CIRCLE) AND HAVE CHILDREN GROUP THEM INTO CATEGORIES USING TECHNOLOGY
- GIVE YOUNG CHILDREN TWO TECHNOLOGY GAMES AND LET THEM EXPLORE THE SIMILARITIES AND DIFFERENCES IN THE GAMES
- PROVIDE TECHNOLOGY ACTIVITIES FOR YOUNG CHILDREN TO IDENTIFY THE OBJECT THAT DOES NOT

BELONG IN A GROUP
- USING THE MOUSE, LET YOUNG CHILDREN IDENTIFY A CERTAIN COLOR OR SHAPE BY POINTING TO IT ON THE SCREEN
- FOLLOWING A TECHNOLOGY ACTIVITY, LET YOUNG CHILDREN LABEL OBJECTS INTO CATEGORIES
- PLAY TECHNOLOGY GAMES WHERE YONG CHILDREN CAN FIND SOME OR ALL OF THE OBJECTS SHOWN ON THE SCREEN
- LET YOUNG CHILDREN IN THE TECHNOLOGY CENTER DESCRIBE A PICTURE ON THE COMPUTER SCREEN
- OBSERVE AND DISCUSS THE MANY DIFFERENT WAYS YOUNG CHILDREN TELL ABOUT THE SAME PICTURE

EARLY CHILDHOOD TECHNOLOGY TEACHING STRATEGY #10

TECHNOLOGY ENRICHMENT EXPERIENCES

The early childhood professional will provide technology enrichment experiences for young children.

TEACHING STRATEGIES TO ENCOURAGE TECHNOLOGY ENRICHMENT EXPERIENCES

- PROVIDE YOUNG CHILDREN WITH NEW TOPICS OF INTEREST FOR THEIR TECHNOLOGY EXPERIENCES
- SHARE NEW PERSPECTIVES TO ENCOURAGE YOUNG CHILDREN TO SEEK NEW TECHNOLOGY INFORMATION
- TRY TO RELATE THE TECHNOLOGY ACTIVITY BACK TO THE THEME AND OTHER CLASS ACTIVITIES
- GIVE YOUNG CHILDREN WHO ARE READY FOR ENRICHMENT GROUP PROBLEMS TO ENCOURAGE WORKING TOGETHER
- LET YOUNG CHILDREN PARTICIPATE IN A VARIETY OF TECHNOLOGY PROJECTS TO PREVENT BOREDOM
- ASK QUESTIONS TO CHALLENGE YOUNG CHILD AND SPUR THEIR INTEREST IN TECHNOLOGY

- MAKE ACTIVITIES FUN AND EXCITING SO YOUNG CHILDREN GAIN A NEW AWARENESS ABOUT TECHNOLOGY
- PROVIDE INTERESTING AND ENRICHING TECHNOLOGICAL ACTIVITIES TO HOLD THE ATTENTION OF YOUNG CHILDREN
- HAVE ESTABLISHED ROUTINES SO YOUNG CHILDREN CAN STAY ON TASK DURING THE TECHNOLOGY ENRICHMENT ACTIVITY
- PLAN PLENTY OF FUN TECHNOLOGY ACTIVITIES FOR YOUNG CHILDREN WHO COMPLETE TASKS QUICKLY

EARLY CHILDHOOD TECHNOLOGY TEACHING STRATEGY #11

TECHNOLOGY SPATIAL RELATIONS EXPERIENCES

The early childhood professional will provide technology opportunities to help young children learn about spatial relations.

TEACHING STRATEGIES TO ENCOURAGE TECHNOLOGY SPATIAL RELATIONS EXPERIENCES

- GIVE STUDENTS OPPORTUNITIES TO TOUCH THE COMPUTER SCREEN TO LOCATE OBJECTS
- LET YOUNG CHILDREN STACK THE DISKS OR COMPUTER SOFTWARE
- AFTER VIEWING PICTURES OF CHILDREN ON A COMPUTER, HAVE EACH CHILD DRAW HIM OR HERSELF
- HAVE YOUNG CHILDREN POINT IN VARIOUS DIRECTIONS ON THE COMPUTER SCREEN
- AFTER LOOKING AT OBJECTS ON A COMPUTER, HAVE YOUNG CHILDREN LOCATE THE SAME OBJECT IN THE CLASSROOM OR LIBRARY
- HAVE YOUNG CHILDREN DESCRIBE DIFFERENT SHAPES USED IN TECHNOLOGY ACTIVITIES

- LET YOUNG CHILDREN MEASURE DISTANCES ON THE COMPUTER SCREEN AND IN HANDS-ON ACTIVITIES
- LET YOUNG CHILDREN VIEW AN OBJECT ON A COMPUTER AND OBSERVE THAT OBJECT IN THE LIBRARY, SCHOOL OR NEIGHBORHOOD
- GIVE YOUNG CHILDREN THE OPPORTUNITY TO OBSERVE THE OBJECT ON A COMPUTER AND THEN TAKE THE ACTUAL OBJECT APART IN A HANDS-ON ACTIVITY
- PROVIDE FOLLOW-UP ACTIVITIES SO YOUNG CHILDREN CAN BUILD THEIR SPATIAL RELATION'S SKILLS

EARLY CHILDHOOD TECHNOLOGY TEACHING STRATEGY #12

TECHNOLOGY SOCIAL EXPERIENCES

The early childhood professional will introduce young children to social experiences through technology.

TEACHING STRATEGIES TO ENCOURAGE TECHNOLOGY SOCIAL EXPERIENCES

- HELP SHY CHILDREN FEEL COMFORTABLE USING TECHNOLOGY IN A GROUP SETTING
- PUT A SHY CHILD WITH A PARTNER DURING A COMPUTER ACTIVITY
- PRAISE A TIMID CHILD FOR COMPLETING A TECHNOLOGY PROJECT WITH ANOTHER CHILD
- ENCOURAGE YOUNG CHILDREN TO SHARE TECHNOLOGICAL EQUIPMENT AND MATERIALS
- ASSIGN SEVERAL CHILDREN CLEAN UP RESPONSIBILITIES FOLLOWING A TECHNOLOGY ACTIVITY
- ENCOURAGE YOUNG CHILDREN TO BE COURTESY TO OTHER PEOPLE WHEN USING TECHNOLOGY
- REINFORCE THAT CHILDREN MUST FOLLOW THE RULES OF THE TECHNOLOGY AREA OR CENTER

- USE SOFTWARE PROGRAMS OR APPLICATIONS TO EMPHASIZE TO CHILDREN THAT PEOPLE CAN HAVE SIMILAR AND DIFFERENT TRAITS
- INTRODUCE YONG CHILDREN TO THE DIFFERENT ROLES, RESPONSIBILITIES AND JOBS OF PEOPLE USING TECHNOLOGY
- USE TECHNOLOGY TO SHOW YOUNG CHILDREN IN SOCIETY THAT PEOPLE DO SOME THINGS ALONE AND SOME THINGS WITH OTHERS

EARLY CHILDHOOD TECHNOLOGY TEACHING STRATEGY #13

TECHNOLOGY TIME EXPERIENCES

The early childhood professional will introduce time to young children during technology experiences.

TEACHING STRATEGIES TO ENCOURAGE TECHNOLOGY TIME EXPERIENCES

- LET YOUNG CHILDREN PUT A STAR ON THE CALENDAR WHEN THEY VISIT THE TECHNOLOGY CENTER
- USE A TIMER AT THE TECHNOLOGY CENTER TO HELP YONG CHILDREN UNDERSTAND DAILY ROUTINES
- PROVIDE YOUNG CHILDREN WITH A START AND STOP SIGNAL SUCH AS A BELL OR NOISE MAKER
- HAVE SOME TIMED TECHNOLOGY ACTIVITIES WHERE YOUNG CHILDREN CAN LEARN ABOUT SPEED (I.E. SLOW AND FAST ACTIVITIES)
- FIND SOFTWARE PROGRAMS THAT WILL HELP YOUNG CHILDREN LEARN ABOUT THE DIFFERENT SEASONS
- DISCUSS AND VERBALIZE TO YOUNG CHILDREN THE PLANS FOR FUTURE TECHNOLOGY EVENTS
- LET YOUNG CHILDREN USE TECHNOLOGY TO EXPLORE

PAST EVENTS AND PARTIES
- PLAN A TECHNOLOGY ACTIVITY WITH YOUNG CHILDREN AND LET THEM HELP PLAN THE SCHEDULE
- PROVIDE YOUNG CHILDREN TECHNOLOGY ACTIVITIES THAT COULD HELP IMPROVE SEQUENCING SKILLS
- FOLLOWING A SOFTWARE PROGRAM, HAVE YOUNG CHILDREN DESCRIBE OR DRAW A PICTURE OF THE PAST EVENT

EARLY CHILDHOOD TECHNOLOGY TEACHING STRATEGY #14

TECHNOLOGY AUDITORY EXPERIENCES

The early childhood professional will help children gain auditory experiences with the use of technology.

TEACHING STRATEGIES TO ENCOURAGE TECHNOLOGY AUDITORY EXPERIENCES

- LET YOUNG CHILDREN LISTEN TO RECORDS AND TAPES WHILE WORKING ON A TECHNOLOGY ACTIVITY
- GIVE YOUNG CHILDREN A CHANCE TO VERBALIZE THEIR IDEAS PRIOR TO A TECHNOLOGY LESSON
- HELP YOUNG CHILDREN BY ORALLY ASKING BRAINSTORMING QUESTIONS DURING A TECHNOLOGY ACTIVITY
- ENCOURAGE YOUNG CHILDREN TO LISTEN TO TECHNOLOGY DIRECTIONS WITHOUT INTERRUPTING
- PROVIDE YOUNG CHILDREN WITH ORAL TECHNOLOGY INSTRUCTIONS
- KEEP ORAL DIRECTIONS SIMPLE WITH ONE OR TWO EASY STEPS OR INSTRUCTIONS
- ASK YOUNG CHILDREN TO REPEAT SOUNDS AND WORDS THEY HEAR FROM THE COMPUTER OR TECHNOLOGY

PROGRAM
- HELP CHILDREN RECALL THEIR FAVORITE POEMS OR RHYMES RELATED TO THE SOFTWARE PROGRAM
- ASK THE YOUNG CHILDREN ORAL QUESTIONS FOLLOWING A TECHNOLOGY LESSON
- INTERVIEW THE CHILDREN ABOUT THEIR DAILY TECHNOLOGY EXPERIENCES

EARLY CHILDHOOD TECHNOLOGY TEACHING STRATEGY #15

TECHNOLOGY LANGUAGE EXPERIENCES

The early childhood professional will provide activities with technology to help your children express and understand language.

TEACHING STRATEGIES TO ENCOURAGE TECHNOLOGY LANGUAGE EXPERIENCES

- HAVE YOUNG CHILDREN VERBALLY IDENTIFY PICTURES ON A COMPUTER SCREEN
- LET YOUNG CHILDREN MAKEUP SENTENCES ABOUT PICTURES THEY SEE ON TECHNOLOGICAL EQUIPMENT
- AFTER SHOWING A STORY WITH TECHNOLOGY, LET THE CHILDREN TELL A STORY THAT HAPPENED TO THEM
- GIVE YOUNG CHILDREN THE OPPORTUNITY TO ORALLY VERBALIZE SOUNDS THEY HEAR FROM THE COMPUTER
- PROVIDE YOUNG CHILDREN WITH TIME TO RETELL A STORY AFTER THEY VIEW IT FROM A SOFTWARE PROGRAM
- HAVE CHILDREN RESPOND TO A STORY BY TELLING ABOUT THE CHARACTERS THEY SEE ON THE MONITOR
- ASK YOUNG CHILDREN QUESTIONS AFTER THEY COMPLETE A SOFTWARE PROGRAM TO CHECK

COMPREHENSION
- LET YOUNG CHILDREN TELL THE DETAILS OF A STORY FOLLOWING THE TECHNOLOGY ACTIVITY
- GIVE YOUNG CHILDREN THE CHANCE TO TELL A STORY IN SEQUENCE AFTER VIEWING A SOFTWARE PROGRAM
- PROVIDE OPPORTUNITIES FOR YOUNG CHILDREN TO INTERACT WITH EACH OTHER WHILE PARTICIPATING IN TECHNOLOGY ACTIVITIES

EARLY CHILDHOOD TECHNOLOGY TEACHING STRATEGY #16

TECHNOLOGY VISUAL EXPERIENCES

The early childhood professional will assist young children in exploring visual experiences through technology.

TEACHING STRATEGIES TO ENCOURAGE TECHNOLOGY VISUAL EXPERIENCES

- ENCOURAGE YOUNG CHILDREN TO USE THEIR EYES AND FOLLOW AN OBJECT ON THE COMPUTER SCREEN
- PROVIDE OPPORTUNITIES FOR YOUNG CHILDREN TO MATCH COLORFUL OBJECTS ON SOFTWARE PROGRAMS
- DISPLAY PICTURES ON THE MONITORS AND HAVE YOUNG CHILDREN LOCATE AND TOUCH OBJECTS
- HAVE YOUNG CHILDREN DESCRIBE HOW OBJECTS LOOK ALIKE AND DIFFERENT FROM SOFTWARE PROGRAMS
- COMBINE VISUAL TECHNOLOGY LEARNING WITH BULLETIN BOARDS, POSTERS AND OTHER VISUAL LEARNING ACTIVITIES
- GIVE CHILDREN TECHNOLOGY OPPORTUNITIES TO VIEW OBJECTS OF DIFFERENT SIZES
- PROVIDE TECHNOLOGY INSTRUCTIONS IN A VISUAL FORMAT TO ASSIST VISUAL LEARNERS

- GIVE YOUNG CHILDREN TIME TO FIND AND IDENTIFY THE MISSING PARTS OF A PICTURE ON A SCREEN
- PROVIDE SOFTWARE PROGRAMS THAT ALLOW YOUNG CHILDREN TO GROUP OBJECTS IN CATEGORIES
- LET YOUNG CHILDREN VISUALLY STUDY A PICTURE ON THE COMPUTER SCREEN AND TRY TO RECALL DETAILS FROM THEIR MEMORY

EARLY CHILDHOOD TECHNOLOGY TEACHING STRATEGY #17

CREATIVE TECHNOLOGY EXPERIENCES

The early childhood professional will provide young children with the opportunity to gain creative technology experiences.

TEACHING STRATEGIES TO ENCOURAGE CREATIVE TECHNOLOGY EXPERIENCES

- HAVE YOUNG CHILDREN CREATE OR DESIGN A PICTURE FROM A SHAPE THEY HAVE VIEWED ON A COMPUTER SCREEN
- PROVIDE WARM-UP ACTIVITIES TO CREATE AN INTEREST ABOUT THE UPCOMING TECHNOLOGY ACTIVITY
- PACE LESSONS SO YOUNG CHILDREN HAVE PLENTY OF TIME TO CREATE NEW THINGS USING TECHNOLOGY
- ENCOURAGE YOUNG CHILDREN TO QUESTION EVENTS SO THEY WILL BECOME MORE CREATIVE WITH TECHNOLOGY
- DISPLAY THE CREATIVE WORK OF YOUNG CHILDREN IN THE TECHNOLOGY AREA OR CENTER
- RELOCATE DISRUPTIVE CHILDREN SO THAT OTHER CHILDREN CAN CREATE AND CONCENTRATE ON THEIR TECHNOLOGY ACTIVITIES

- BE SENSITIVE WHEN YOUNG CHILDREN BECOME FRUSTRATED WITH TECHNOLOGY ACTIVITIES THAT STRESS CREATIVITY
- GIVE YOUNG CHILDREN PLENTY OF TIME TO FINISH A CREATIVE TECHNOLOGY PROJECT (IT MAY TAKE LONGER FOR SOME CHILDREN)
- LET YOUNG CHILDREN CREATE A SKIT WITH PROPS AFTER VIEWING A STORY DURING TECHNOLOGY TIME
- SHOW A FUNNY VIDEO AND ALLOW YOUNG CHILDREN TO MAKE UP FUNNY EXPRESSIONS OR SILLY ACTIONS RELATED TO THE VIDEO

EARLY CHILDHOOD TECHNOLOGY TEACHING STRATEGY #18

TECHNOLOGY EMOTION EXPERIENCES

The early childhood professional will give young children the opportunity to learn about emotions and feelings with technology.

TEACHING STRATEGIES TO ENCOURAGE TECHNOLOGY EMOTION EXPERIENCES

- USE THE COMPUTER TO SHOW YOUNG CHILDREN THE DIFFERENT FACIAL EXPRESSIONS THAT PEOPLE EXHIBIT
- INCORPORATE FINGER PLAYS WITH COMPUTER SOFTWARE PROGRAMS SO SHOW DIFFERENT EMOTIONS
- LET CHILDREN EXPRESS THEIR FEELINGS THROUGH PLAY ACTIVITIES FOLLOWING A TECHNOLOGY LESSON
- DISCUSS HOW CHARACTERS FEEL AFTER YOUNG CHILDREN VIEW NURSERY RHYMES THROUGH TECHNOLOGY
- USE POEMS THAT GO ALONG WITH COMPUTER SOFTWARE PROGRAMS TO REINFORCE DIFFERENT FEELINGS
- LET YOUNG CHILDREN DANCE TO EXPRESS EMOTIONS THAT WERE INTRODUCED THROUGH TECHNOLOGY

- PROVIDE OPPORTUNITIES FOR YOUNG CHILDREN TO MAKE FACES IN A MIRROR TO IMITATE AN IMAGE ON THE COMPUTER
- PROVIDE DRESS UP CLOTHES FOR CHILDREN TO ROLE PLAY THE EVENT
- ENCOURAGE YOUNG CHILDREN TO LISTEN AND OBSERVE THE EMOTIONS OF CHARACTERS ON A SHORT VIDEO
- HAVE CHILDREN DISCUSS THEIR OWN FEELINGS AND RELATE THEM TO A CHARACTER ON A VIDEO

EARLY CHILDHOOD TECHNOLOGY TEACHING STRATEGY #19

TECHNOLOGY MUSIC EXPERIENCES

The early childhood professional will provide technology experiences for young children to learn about music.

TEACHING STRATEGIES TO ENCOURAGE TECHNOLOGY MUSIC EXPERIENCES

- EMPHASIZE TO YOUNG CHILDREN THE IMPORTANCE OF LISTENING TO MUSIC AND SOUNDS COMING FROM THE COMPUTER
- PLAY MUSICAL GAMES DURING TECHNOLOGY ACTIVITIES
- ALLOW YOUNG CHILDREN TO DANCE TO MUSIC GENERATED THROUGH TECHNOLOGY
- LET YOUNG CHILDREN PARTICIPATE IN CREATIVE MOVEMENT EXERCISES ALLOWING MUSIC TECHNOLOGY ACTIVITIES
- GIVE YOUNG CHILDREN A CHANCE TO EXPLORE A MUSICAL INSTRUMENT AFTER STUDYING A CULTURE ON THE COMPUTER
- ENCOURAGE YOUNG CHILDREN TO SING WITH A COMPUTER SOFTWARE PROGRAM

- HAVE ALL YOUNG CHILDREN PARTICIPATE IN A MUSIC TECHNOLOGY ACTIVITY BY PLAYING INSTRUMENTS WITH AN APPROPRIATE MUSICAL SOFTWARE PROGRAM
- USE A HAND CLAPPING OR TAPPING EXERCISE TO KEEP THE BEAT OF THE MUSIC DURING A TECHNOLOGY SOFTWARE PROGRAM
- PLAN TECHNOLOGY MUSIC ACTIVITIES RELATED TO SPECIAL THEMES, HOLIDAYS AND EVENTS
- HAVE YOUNG CHILDREN TRY TO IDENTIFY CERTAIN SOUNDS AND NOISES THEY HAVE HEARD DURING A MUSIC TECHNOLOGY ACTIVITY

EARLY CHILDHOOD TECHNOLOGY TEACHING STRATEGY #20

POSITIVE TECHNOLOGY LEARNING EXPERIENCES

The early childhood professional will help children gain positive learning experiences using technology.

TEACHING STRATEGIES TO ENCOURAGE POSITIVE TECHNOLOGY LEARNING EXPERIENCES

- COMMUNICATE TO YOUNG CHILDREN WITH SMILES AND POSITIVE GESTURES AS THEY WORK WITH TECHNOLOGY
- COMPLIMENT YOUNG CHILDREN FOR TAKING CARE OF THEIR OWN NEEDS WHEN USING TECHNOLOGY
- PRAISE YOUNG CHILDREN OFTEN FOR THE STEPS THEY MAKE WHEN USING TECHNOLOGICAL EQUIPMENT
- COMMEND YOUNG CHILDREN AS THEY LEARN NEW SKILLS FOR WORKING WITH TECHNOLOGY
- PROVIDE POSITIVE COMMENTS FOR YOUNG CHILDREN AS THEY SELECT AND MAKE CHOICES FROM SOFTWARE PROGRAMS
- REWARD YOUNG CHILDREN FOR WORKING TOGETHER

COOPERATIVELY ON TECHNOLOGY ACTIVITIES
- DISCUSS MEANINGFUL TECHNOLOGY EXPERIENCES WITH YOUNG CHILDREN
- PRAISE BOTH GROUPS AND INDIVIDUALS FOR THEIR EFFORTS ON TECHNOLOGY ACTIVITIES
- BRING PARENTS TO LIBRARY AND SCHOOL ACTIVITIES TO REINFORCE POSITIVE TECHNOLOGY EXPERIENCES
- PROVIDE STICKERS AND AWARDS FOR YOUNG CHILDREN TO PUT ON THEIR COMPLETED TECHNOLOGY ACTIVITIES

CHAPTER 3

EARLY CHILDHOOD TECHNOLOGY REWARDS

Young children taking on any new learning endeavor will always need encouragement, praise and motivation to meet the challenges of the learning process. This chapter focuses on many areas related to rewarding learners in early childhood technology. The topics covered in this chapter include:

WHY PRAISING TECHNOLOGY USE IS IMPORTANT?
IDEAS FOR PRAISING TECHNOLOGY USE
TECHNOLOGY PRAISE PHRASES
MOTIVATION TIPS FOR EARLY LEARNERS USING TECHNOLOGY
SIMPLE COMPUTER TASKS FOR EARLY LEARNERS
TEACHERS AND LIBRARIANS CAN MODEL POSITIVE TECHNOLOGY USE
BUILDING A CHILD'S SELF-ESTEEM WITH TECHNOLOGY
A TECHNOLOGY REWARD SYSTEM FOR EARLY LEARNERS

WHY PRAISING TECHNOLOGY USE IS IMPORTANT

Young children need praise and encouragement when using technological equipment because:

- it will give children positive technology experiences
- it will provide children support when they make mistakes
- it will give children courage to work on new technology devices
- it will build the self-confidence of children in using computers
- it will help children be brave in trying new things
- it will encourage children to work in teams and cooperate
- it will give children confidence to ask questions and seek help
- it will help children in exploring new situations
- it will assist children in gaining assurance about their abilities
- it will give children support as they face new challenges

IDEAS FOR PRAISING TECHNOGY USE

Young children can be praised for many technology-related tasks:

turning on or off a computer or technological device
remembering a previously taught skill
asking permission to use technological equipment
helping a classmate with a technology problem
giving technology directions to a new student
working quietly on a computer
using manners when working on technological equipment
sharing technological devices with others
finishing a difficult or challenging technology task
completing technology assignments and tasks

TECHNOLOGY PRAISE PHRASES

The following phrases could be used to praise the efforts of young children using **technology:**

- "I like the way you turned on the computer."
- "The computer group is working quietly."
- "Thank you for waiting your turn to use the computer."
- "I liked the way you helped Sophie at the computer."
- "The computer group lined up very quietly today."
- "Everyone in the computer group finished their work today."
- "The technology group cleaned up their center fast today!"
- "I like the way the leader helped the technology group."
- "Everyone was smiling at the computer center today!"
- "I appreciate the way the computer group is completing their work."

MOTIVATION TIPS FOR EARLY LEARNERS USING TECHNOLOGY

Motivating young children to use technology can take many forms:

- MAKING TECHNOLOGY PROGRAMS EXCITING
- COMMENDING CHILDREN'S TECHNOLOGY EFFORTS
- GEARING TECHNOLOGY INSTRUCTION TO THE NEEDS OF CHILDREN
- GIVING CHILDREN OPPORTUNITIES TO BE LEADERS
- ALLOWING BRAINSTORMING TECHNOLOGY ACTIVITIES
- PRAISING CHILDREN'S TECHNOLOGY QUESTIONS
- REWARDING CHILDREN'S ACCOMPLISHMENTS
- PROVIDING STIMULATING TECHNOLOGY ACTIVITIES
- REDIRECTING NEGATIVE COMMENTS INTO POSITIVE STATEMENTS
- ASSISTING STUDENTS HAVING DIFFICULTY IN TECHNOLOGY TASKS

SIMPLE COMPUTER TASKS FOR EARLY LEARNERS

The following information is a list of simple computer tasks that could be used to build the self-confidence of young children:

- Find the disk with the red circle on it.
- Press the button with the blue dot on it
- Touch the monitor.
- Press a button on the keyboard.
- Put your hand on the mouse.
- Touch the mouse pad.
- Pick up a disk.
- Touch the printer.
- Place your finger on the screen.
- Pick up the computer paper.

WAYS TEACHERS AND LIBRARIANS CAN MODEL POSITIVE TECHNOLOGY USE

Teachers and librarians have many opportunities to model positive technology experiences by their actions and behaviors. The following is a list of some of these modeling behaviors:

- Keep relaxed even when children make mistakes on the computer
- Be sensitive when children cry or seem frustrated from using technology
- Listen attentively to children's technology questions and concerns
- Smile and be enthusiastic about technology
- Demonstrating technology with simple and easy instructions
- Helping children feel comfortable with technological equipment
- Giving children directions with a friendly attitude
- Being helpful when children are involved in group and individual technology activities
- Offering assistance to young children who need extra technological support

BUILDING A CHILD'S SELF ESTEEM WITH TECHNOLOGY

Teachers, librarians and early childhood professionals have many opportunities to build a child's self esteem during technology learning projects and activities. A few examples are listed below:

- Helping a young child set a technology goal
- Assisting a child in finding technology solutions
- Brainstorming technology topics of interest with a child
- Planning technology discovery activities for children
- Identifying the strengths of a child using technological equipment
- Sharing successful technology experiences with young children
- Encouraging young children to be good sports when playing technology games
- Praising group and individual efforts when using educational software programs
- Setting up challenging technology activities to enhance and enrich learning

A TECHNOLOGY REWARD SYSTEM FOR EARLY LEARNERS

Early learners who are rewarded for technology use may become motivated to reach new challenges in their technological experiences. A technology reward system may take many forms, but here are some suggestions:

- GIVING YOUNG CHILDREN OPPORTUNITIES TO BE TECHNOLOGY LEADERS AND EXPERTS FOR A DAY
- LETTING A CHILD BE THE TEACHER'S TECHNOLOGY HELPER FOR A WEEK
- ALLOWING A CHILD TO PASS OUT TECHNOLOGY STICKERS AT THE END OF A TECHNOLOGY LESSON
- PROVIDING FREE PLAY ON THE COMPUTER
- CHOOSING THE EDUCATIONAL SOFTWARE PROGRAM OR GAME FOR THE CLASS
- SELECTING A PARTNER OR TEAM FOR THE TECHNOLOGY CENTER
- PICKING A CHILD TO HAND OUT TECHNOLOGY REWARDS AT THE END OF THE WEEK
- MAKING A BULLETIN BOARD TO DISPLAY TECHNOLOGY ACCOMPLISHMENTS
- RECOGNIZING A TECHNOLOGY STUDENT OF THE WEEK
- REWARDING TECHNOLOGY EFFORTS WITH A MONTHLY COMPUTER PARTY WHERE SOFTWARE GAMES AND TREATS ARE PROVIDED

CHAPTER 4

SAFETY AND EARLY CHILDHOOD TECHNOLOGY

Teachers and librarians working with young children in early childhood technology programs will face a variety of questions and issues about safety. The topic of supervision will also be of importance as teachers and librarians work with young children and technological equipment. The goal is to have a safe and responsible technology learning environment for young children. Some of the issues covered in this chapter include:

WHY TECHNOLOGY SUPERVISION IS NEEDED
SUPERVISORY HELP FOR EARLY CHILDHOOD TECHNOLOGY
TECHNOLOGY AREAS THAT MAY CAUSE ACCIDENTS
PROFILES OF EARLY TECHNOLOGY LEARNERS
RULES FOR AN EARLY CHILDHOOD TECHNOLOGY CENTER
SAFETY CAUTIONS FOR AN EARLY CHILDHOOD TECHNOLOGY AREA
WAYS TO ENCOURAGE TECHNOLOGY SAFETY FOR YOUNG CHILDREN
WAYS TO PREVENT EARLY CHILDHOOD TECHNOLOGY ACCIDENTS

WHY TECHNOLOGY SUPERVISION IS NEEDED

Technology supervision for young children is essential when establishing an early childhood technology program. Technology supervision is needed to:

- Protect the health of young children
- Prevent accidents with technological equipment
- Keep children away from dangerous technology situations
- Stop children from damaging technological equipment
- Hinder children from potential injuries
- Avoid harming other children in the class or technology center
- Redirect children from unsafe technology areas
- Help children out of unsure technology situations
- Alert children of actions that might harm them
- Check equipment for hazardous trouble spots

SUPERVISORY HELP FOR EARLY CHILDHOOD TECHNOLOGY

There are many people around a school, day care center or library that could help in a supervisory capacity when young children are working with technological equipment. Some of the people that might be of assistance include:

- Teachers
- Librarians
- Instructional Assistants
- Library Graduate Students
- Interns
- Parents
- School and Library Volunteers
- College students in practicum classes
- Student teachers
- Job training apprentice

TECHNOLOGY AREAS THAT MIGHT CAUSE ACCIDENTS

Teachers and librarians may want to assess their technology areas to make sure they are safe. Some of the areas to observe include:

- Equipment Arrangement
- Partitions
- Cords
- Electrical Outlets
- Storage Units
- Shelves and Drawers
- Dividers/Room Organizers
- Tables
- Supply Storage
- Cabinets

PROFILES OF EARLY TECHNOLOGY LEARNERS

Young children in an early childhood technology center will have many different personality and behavior profiles. Some of these profiles include:

Children with various opinions.
Children with different ability levels.
Children with inconsistent behavior.
Children with a variety of technological interest levels.
Children with mood swings and emotional concerns.
Children with varying self-confidence levels.
Children with individual needs.
Children with varying degrees of life experience.
Children with different types of language experiences.
Children with a wide range of physical capabilities.

RULES FOR EARLY CHILDHOOD TECHNOLOGY CENTERS

Each early childhood technology center will be different and may need its own guidelines and rules. Here are some general rules that early childhood professionals might want to consider in an early childhood technology center:

- Walk in the technology area
- Wash hands before using the computers
- Keep pencils, crayons, scissors and glue in cubbyholes (these items can damage computers)
- Work quietly (or use soft voices) in the technology center
- Be careful not to bang or scratch the computer
- Remember computers and equipment are not toys, so be careful when using them
- Keep snacks and drinks away from the computer and equipment
- Use good manners and say kind words when working in the technology center
- Push in the chairs after you are finished with the computers
- Clean up the garbage in the technology center and take it to the trash can

SAFETY CAUTIONS FOR AN EARLY CHILDHOOD TECHNOLOGY AREA

Young children have unique experiences and personalities so there are some safety cautions professionals might want to consider:

- Remove small computer pieces that children may swallow.
- Watch out for sharp points or edges children may bump.
- Place long cords out of the children's path so they will not trip.
- Place equipment-cleaning supplies out of children's reach.
- Cover electrical outlets that are not in use.
- Move equipment from the edge of tables to prevent falling off of tables.
- Watch areas closely because children can get their little fingers and hands caught inside of equipment.
- Be careful of long hair and braids that could get caught in technological equipment.
- Make sure shelves are sturdy to prevent items from falling off high places.
- Be aware of clothing that might damage or get caught in technology equipment.

WAYS TO ENCOURAGE TECHNOLOGY SAFETY FOR YOUNG CHILDREN

Some ways to encourage technology safety for young children include:

- Review technology rules and guidelines daily
- Design a technology safety bulletin board
- Send notes home to parents and invite them to be part of the technology safety effort
- Have a technology safety open house
- Hand out stickers for children who show good safety habits with technological equipment
- Make technology safety projects and share them with other classes and groups
- Praise the efforts of young children as they learn the technology safety rules
- Recognize outstanding technology safety with certificates and awards
- Have a play and let children dramatize technology safety rules
- Show a technology safety video or sing safety songs

WAYS TO PREVENT EARLY CHILDHOOD TECHNOLOGY SAFETY PROBLEMS

Here are some ways to prevent early childhood technology accidents:

- Encourage cleanliness when using computers to prevent the spread of diseases
- Supervise young children at all times when using technological equipment
- Provide routine maintenance for computer and technological equipment
- Check technological equipment for unsafe parts that could pinch or cut fingers and hands
- Make sure harmful supplies are stored out of children's view and reach
- Review rules before a technology lesson and when children forget them
- Send a note home to parents and ask them to help their children learn about technology safety
- Plan age appropriate technology activities to prevent children from becoming frustrated and angry with technology
- Make sure tables, shelves, cabinets and storage areas are securely fastened to prevent equipment and supplies from falling
- Ask for volunteers and extra personnel to help supervise technology groups when necessary

CHAPTER 5

QUESTIONS ABOUT EARLY CHILDHOOD TECHNOLOGY

Early childhood professionals are not always included in the school district and library technology training programs. Some school districts have the attitude that young children do not need technology education and that they will get it later in the upper grades. There are some early childhood programs that really emphasize the importance of technology for young children, while others never focus their programs toward technological experiences for young children. As a result, children coming to library programs and schools will have a wide variety of technological experiences. Some children will have confidence and excitement about computers and technology, while other children will be unsure of themselves and their abilities when they confront technological situations.

Early childhood professionals will also have a variety of experiences, education and work training in technology areas. Their different ability levels and confidence will reflect in their teaching of technology to young children. The fast pace of technological developments have left many early childhood education professionals feelings barraged with mountains of technological information and no time to gear it to the level of early childhood children. The technological advances do not always address specific interests and needs of young children. Early childhood teachers have a number of questions and concerns about technology and its usefulness in the school and library early childhood programs. School districts, libraries and early learning programs may want to consider some of these questions as they design future technology programs and update established technology procedures.

- How are early childhood professionals involved in the software selection process?
- Are parents involved in school and library early childhood technology programs?
- Does the school have an early childhood technology program?
- What is the library software budget for early childhood education?
- At what grade level do students begin to receive computer training?
- Are there funds for early childhood technology equipment and supplies?
- How can early childhood programs be modified to include technology?
- Is there specific technology training geared to assist professionals in early childhood education?
- Who oversees the school and library early childhood technology programs?
- Is there a schedule of technology training for librarians working with young children?

Questions About Early Childhood Technology

- What early childhood technology games are available to professionals?
- How can technology be used to reinforce early learning?
- Is there a catalog of district approved software for early childhood students?
- Are there technology tutorial programs for young children?
- How can technology be used to promote reading readiness?
- What types of college courses are available in early childhood technology?
- Are there district guidelines for using early childhood technology?
- How can early reading and storytelling be enhanced by technology?
- Who is responsible for developing early learning technology programs?
- How do parents and teachers get crayon marks off a computer?
- Can technology be used to enhance early childhood field trips?
- What is the school's philosophy of early childhood technology?
- Are there technology workshops for early childhood teachers?
- How can technology be adapted to a learning center approach?
- What are some ways to design early childhood technology activity centers?
- How can technology be used in a developmentally appropriate way with young children?
- What are some ways technology can promote peer-interaction in young children?
- What are some ways to use technology in small group activities?

- How can technology be combined with a variety of teaching methods?
- Are there ways to use technology with a whole group approach?
- What are some of the strengths of using technology in early childhood?
- What are some realistic technology goals for young children?
- How can a young child's communication skills be enhanced by technology?
- What types of technology opportunities exist for the young child?
- How can technology be used in play activities?
- Can there be a hands-on approach in all technology activities?
- Is there any information on multicultural technology for young children?
- How does technology enhance a young child's decision-making skills?
- What questioning techniques can be used with early childhood technology?
- Can block activities be used with early childhood technology?
- Do young children find independence from technology activities?
- Where can early childhood teachers find technology materials?
- Does technology help a young child's self discipline?
- Is it possible for young children to select their own technology activities?
- Can technology be used effectively with pre kindergarten students?

Questions About Early Childhood Technology

- Does technology address young children at different levels?
- How is creativity promoted from early childhood technology?
- Should technology be teacher initiated or student initiated for young children?
- Is a young child's confidence increased from technology use?
- How does technology allow young children to see different viewpoints?
- Are there suggestions for classroom arrangements using technology?
- Do technology programs address young children with special needs?
- What are some rules for using early childhood technology?
- Do the technology activities provide positive experiences for young children?
- Is there technology software for language development?
- Can young children design things using technology?
- Are social relations enhanced by early childhood technology?
- Can technology help young children improve physical development?
- What are some technology activities to enhance free exploration?
- How can technology be related to discovery learning?
- How can technology be combined with dramatic play?
- Is a child's social interaction limited with the use of technology?
- Where do early childhood teachers store technology equipment and materials?
- How does technology fit into the early childhood daily schedule?

Early Childhood Technology Planning

- Is technology exciting for young children?
- What are some ways to evaluate early childhood technology?
- Wow can science for young children be used with technology?
- What are some ways to improve a child's self esteem with the use of technology?
- Is eye-hand coordination improved with technology use?
- How can technology be used in fine motor development?
- What is some safety measures in using technology with young people?
- Can classroom art activities be created with technology?
- When is it appropriate to use props with technological equipment?
- How can technology be used to improve a child's listening skills?
- What types of technology projects are geared toward early learning?
- How does technology fit into the curriculum goals of early childhood?
- What is the rationale behind using technology with early childhood students?
- How can long books be used with technological equipment?
- Can technology help young child interpret stories?
- What are some ways parents can be involved in early childhood technology?
- How can technology be combined with role-playing activities?
- Can technology help young children distinguish things that are real and make believe?

Questions About Early Childhood Technology

- Are computers stressful for young children?
- How can young children predict outcomes with technology?
- Do young children get eyestrain from using computers?
- Can technology activities be sequenced for young children?
- What are some ways to use technology with early math activities?
- How can children use technology for patterning?
- Can young children's counting skills be improved with technology?
- How can technology be used to teach young children about occupations?
- Can sorting activities be used with classroom technology?
- What types of questions will increase curiosity in early learners?
- How can technology be used to intellectually challenge young children?
- What early childhood teaching strategies can be used with technology?
- Will technology frustrate young children?
- What are some ways early childhood teachers can grade technology activities?
- How can children help each other in technology activities?
- What technology skills will help encourage future learning for young children?
- Can young children identify details using technology?
- What are the expected outcomes of using early childhood technology?

Early Childhood Technology Planning

- Is there an inventory of early childhood technology equipment?
- How can technology help the parents of young children?
- Who provides technology leadership for early childhood teachers?
- What does current research suggest about technology and young learners?
- Have the early childhood teachers been surveyed on their technological experiences?
- Does technology address young children with a wide range of ability levels?
- What technology responsibilities are required of early childhood teachers?
- Are there technology unit plans developed for early childhood teachers?
- How can individual instruction be used with classroom technology?
- What way does technology interact with a child's sensory awareness?
- I there technology orientation for early childhood teachers?
- How does technology impact the instructor's teaching style?
- Can technology be used to help children establish routines?
- How can young children be rewarded for technology use?
- What types of choices can children make using technology?
- Are there color activities that can be completed on the computer?
- How can children learn to follow directions using a computer?
- Can children learn sounds with a computer?

Questions About Early Childhood Technology

- What way does technologies impact a child's self-concept?
- How can technology increase a young child's vocabulary?
- Can shapes be taught using a computer?
- How can the alphabet be emphasized with technology?
- What are some counting activities using technology?
- Is there any software to teach young children about money?
- How can children retell stories with technology?
- What are some good habits students can learn from technology?
- Where can early childhood teachers find technological information?
- Will using technology put high pressure on young children?
- What are some early childhood teacher expectations for using technology in the classroom?
- Is there a way to use technology to learn about community workers?
- Can young children work in groups using one computer?
- Is there software to help young children learn to follow directions?
- How can young children learn about the work of using technology?
- How can children learn about feelings using technology?
- What are some courteous things children can do on computers?
- Can young children learn cultural information using computers?
- Are there software programs to teach children about family life?
- Does using technology limit opportunities for children to speak

and use oral language?
- Can finger plays be used with computer programs?
- How can a calendar be used with technology?
- Is there a software program for students to compare the size of objects?
- Can graphs be computer generated for students to organize information?
- How can technology be combined with daily opening activities?
- Where does technology fit into the thematic unit approach?
- How can children better express themselves using technology?
- What is some ways technology can be used with oral presentations?
- Can dramatic play be enriched from using technological equipment?
- When can technology be used to correct mistakes in the classroom?
- What types of creative movement activities can be used with technology?
- What ways can technology be used to teach transportation?
- What are some ways computers can be combined with books?
- Does the school climate encourage early childhood technology?
- Is there a school planning committee for early childhood technology?
- Does the district have any long-range plans for technology in early childhood education?
- How should technology be introduced to preschoolers?
- Is there a division in charge of early childhood technology?

Questions About Early Childhood Technology

- What is the focus of early childhood technology in the school district?
- What school personnel members are responsible for early childhood technology?
- Who implements future school technology planning?
- Is each school responsible for its own early childhood technology?
- Are grants being developed to fund early childhood technology?
- What community resources might help fund early childhood technology?
- Can business partnerships be formed to add computers in early childhood classes?
- Who handles early childhood technology complaints?
- Who is a good contact person for an early childhood technology committee?
- Are there qualified personnel to teach early childhood technology?
- Does early childhood technology fit into the school district's mission?
- Who could develop an early childhood technology handbook?
- Is there an early childhood technology newsletter?
- Could early childhood technology advisory group be formed?
- Who could plan an open house for early childhood technology?
- What organization could plan an early childhood technology conference?
- How can parents be encouraged to volunteer in early childhood technology?

- Is there extra training for early childhood teachers to learn technology?
- Are there teaching materials for early childhood technology?
- How can technology be labeled in the room?
- Where can teachers share their opinions about early childhood technology?
- How can bulletin boards be used to reflect early childhood technology?
- Should early learners be in various groups to use computers?
- Who could develop press releases to promote early childhood technology?
- Could team teaching be used for early childhood technology?
- How can early learning students explore their own interests using technology?
- What early childhood topics can be covered effectively with technology?
- How can students be encouraged to take challenges when using technology?
- What technology equipment fits well early childhood classrooms?
- Should early learners have a partner when using a computer?
- Can puppets be used effectively with early childhood technology?
- Do early childhood teachers feel competent to teach technology?
- Sound a timer be used for early childhood technology activities?
- How can early learners feel secure using technology in the classroom?

Questions About Early Childhood Technology

- What types of flannel board activities can be combined with technology?
- Coined there be a technology-training program for the parents of early learners?
- How can early childhood teachers encourage apprehensive children to learn technology?
- What are some ways to design an early childhood technology center in a classroom?
- How can early childhood teachers record the technology progress of their students?
- What are some ways to use technology to promote a child's imagination? How can early childhood teachers effectively demonstrate technology skills to their peers?
- How can early childhood teachers effectively demonstrate technology skills to their peers?
- How can early childhood technology be used to help problem students?
- Where can teachers find information on model early childhood technology programs?
- How can early learners be motivated to use technology?
- Can listening skills be facilitated with the use of technology?
- Does technology limit oral language activities?
- How can technology help students with physical education?
- What kind of theme units can be created with technology?
- Can charts and patterns be used with technology?
- What type of artwork can be created with computers?
- Are there early childhood technology planning sessions?

- Is there an early childhood technology resource guide for new teachers?
- How can technology be integrated into an early childhood curriculum?
- Can children make connections to writing using technology?
- Is there a way to provide story extension with technological equipment?
- Are there opportunities to reinforce academic skills with technology?
- Is there a way to use a computer with rhyming words?
- Can young children learn about sounds with technological equipment?
- How can a story be introduced with a computer?
- Can technology be used in connection with nursery rhymes?
- What are some ways technologies could be used to encourage curiosity?
- Can technology be part of transitional activities in a classroom?
- Is there a way to introduce travel to young children using technology?
- How can young children use technology to investigate letters?
- Where can early childhood teachers find developmentally appropriate technology activities?
- What are some ways an instructional assistant can help with technology activities?
- Can technology supplement the present early childhood curriculum?
- What are some ways to teach math skills with a computer?

Questions About Early Childhood Technology

- How can we link technology to cultural experiences?
- Can food activities be used with technology?
- Is there a bibliography of early childhood technology resources?
- Can young children improve prediction skills with technology?
- How can enthusiasm be created for young children and technology?
- What are some dance activities to use with technology?
- Can finger plays be enriched with technology?
- What are some writing and drawing skills that can be emphasized with teaching?
- Can technology be used to help young children with social skills?
- What are some ways to sue technology to help young children with disabilities?
- What research has been conducted on early childhood technology?
- Can older children help young children with technology skills?
- Should young children go to the school's technology lab?
- Can technology be used to enhance comprehension skills?
- What skills do young children need to use a computer?
- What are some good resources for teaching technology in early childhood?
- How flexible are early childhood technology centers?
- What does technology use do to promote eye-hand coordination?
- Does technology help young children in motor development?
- Can concepts of print be taught on a computer?
- What background knowledge is needed to run technological

equipment?
- Does technology help build the vocabulary of young learners?
- How could young children play pretend games and use technology?
- Can technology be used to sequence events?
- What type of attitudes do young children have toward technology?
- Can young children learn about real and imaginary things on a computer?
- How can computer illustrations be presented to young learners?
- Can pantomiming be incorporated with technology?
- What types of props can be used with technology?
- What are some ideas for relating computers to a theme?
- How can the computer be used to generate science activities for early learners?
- Can lessons be introduced to young children with a computer?
- How can a technology center be personalized for young children?
- Are there technology activities to use in a variety of learning centers?
- What are the assessment issues for early childhood technology?
- Does the room have child-sized tables and chairs to hold computers and technology equipment?
- What are some pre-lesson motivators to introduce technology lessons to young children?
- Where can teachers find technology materials for a space

Questions About Early Childhood Technology

lesson?
- Is it possible to retell a story using a computer?
- Are there stories and events for young children to sequence on a computer?
- Can young children use technology to help recall characters in a story?
- Is there enough time for young children to have repeated opportunities on technological equipment?
- Are there teaching activities for young children to learn letters in the alphabet?
- Could a young child keep a writing journal on a compute4r?
- What are some ways to encourage technological explorations in young children?
- Can young children use technology at their own pace?
- Does a child have an opportunity to learn directionality with computer activities?
- Could a young child use a computer for a show and tell project?
- Is technology useful in helping children with matching skills?
- How can punctuation be taught with technology?
- Can early learners use one to one correspondence with a computer?
- What are some ways an early childhood teacher can model technology use?
- Is there a way to use newspapers with technology activities?
- How much daily class time should be spent on computers and technology?
- Can poems be used effectively with technology activities?

Early Childhood Technology Planning

- What are some things young children can construct on a computer?
- How could vocabulary words be presented to children with technology?
- How many days a week should technology lessons be presented?
- Should young children preview a lesson on the computer?
- What are some ways to teach language arts with technology?
- Can technology be used with circle time activities?
- What are some ways young children can learn to take turns on a computer?
- Is there a way children can learn about symbols with technology?
- Can technology help increase the attention span of young children?
- Does using technological equipment hurt a child's posture?
- How can technology be used to highlight features of a lesson?
- What are some early childhood teacher-prompts for technology activities?
- Can children read books from the computer screen?
- How can a computer be used to produce a play for children?
- Can a video camera or cassette player be part of a technology center?
- What type of equipment could be part of a technology learning center?
- Who are some early childhood technology resource people in the school district?
- What are some interesting library activities to do with

technology?
- How can young children better appreciate literature by using technology?
- What types of art and color activities can be created with technology?
- How can technology make school activities fun for young children?

RECOMMENDED READING

Peterson, S. (2015). *Digital Research Phrase Book*

The **Digital Research Phrase Book** contains thousands of research phrases from a digital and electronic perspective. Many phrases in research reports and papers often have a technological or computer type of focus. The book lists phrases and examines developing ideas and phrases from a virtual perspective of looking into network relationships, electronic features, technology aspects, online data collection and various cyber elements that connect the research and testing world. This valuable digital research book would compliment libraries, graduate and undergraduate programs, writing centers and departments expanding to more of a technological focus. College students and faculty involved in developing research proposals, thesis ideas, college presentations, grant proposal requests and academic papers can benefit from using these phrases to spark research ideas with a technological and digital focus.

Peterson, S. (2015). *Tech Research Phrase Book*

The *Tech Research Phrase Book* helps focus many of the technology elements toward writing research focused essays, proposals and reports. A simple technology phrase can be easily transformed and developed into longer sentences and paragraphs to describe a research based concept. In this book, technology phrases can be used to develop a focus or objectives for a research study or grant. As well, the technology phrases can be helpful in developing the investigation, assessment and evaluation components of a paper or proposal. These technology phrases can be used to describe results and establish recommendations for a project. Technological change phrases are included in the book as an ongoing feature of the changing nature of technology and how it impacts research and writing projects. The ***Tech Research Phrase Book*** definitely provides a beginning place to develop technology phrases into effective research report writing.

INDEX

A

Accident (s), 71
Application (s), 5
Apprentice, 65
Art, 11, 23-24
Auditory, 12, 39-40

B

Behavior, 67

C

Camp (s), 9
Capabilities, 66
Challenge (s), 4, 40
Classification, 11, 29-30
Clothing, 69
Computer (s), 68
Creative, 45-46

D

Damage, 69
Discovery, 11, 25-26

E

Educational materials, 6

Emotion, 12, 47-48

Enrichment, 8, 12, 31-32

Equipment, 41, 54, 64

Experiences, 43-44

F

Family, 9

Fine motor, 11, 15-16

Finger (s), 69

H

Hair, 6

Health, 49

I

Imagination, 69

Individual, 52

Information, 6, 25

Injuries, 64

Internet, 65

J

Jumping jacks, 21

Index

L
Language, 12, 41-42
Librarian (s), 8
Life experiences, 52

M
Monitor, 44 96
Motivation, 57
Music, 12, 49-50

N
Needs, 52

O
Objective (s), 3
Online, 13
Open house, 70
Opinion (s), 67
Outlet (s), 66, 69

P
Parent (s), 9, 65
Planning, 11, 13
Positive learning, 12, 51-52
Praise, 53
Professional (s), 17
Puppet (s), 68

Q
Question (s), 33, 57

R
Rationale, 2
Reading, 16
Resources, 5
Responsibility, 21-22
Reward (s), 56
Role Play, 11, 27-28
Rules, 48

S
Safety, 63, 69
Self-Esteem, 60
Sensory, 19-20
Sequence, 33
Sightseeing, 14
Social, 12, 35-36
Software, 70
Songs, 55
Spatial, 12, 33
Supervision, 64

T
Teacher (s), 7
Team (s), 61

Index

Technology, 59
Time, 37-38

U
Unit plans, 64

V
Visual, 43-44
Vocabulary, 64, 98
Volunteer (s), 5, 56

W
Webcamp (s), 5
Websites, 5
Workshop (s), 5-6

Y
Yearly, 13

AFTERWORD

Early Childhood Technology Planning was written to encourage more planning in the area of technology for young children. Children are currently faced with more technology decisions than their parents. Some young children are blessed with tech savvy parents who provide the latest equipment, but other children from at-risk situations may not even have a computer and have limited exposure to technology in the home. Schools and libraries can provide this technological support through early childhood technology planning. **Early Childhood Technology Planning** is a wonderful initial planning resource for early childhood professionals, librarians and staff to include the much needed technology component in early childhood programs.

www.ingramcontent.com/pod-product-compliance
Lightning Source LLC
Chambersburg PA
CBHW021131300426
44113CB00006B/388